Original Korean text by Ki-hwa Jang
Illustrations by Seung-min Oh
Korean edition © Aram Publishing

This English edition published by big & SMALL in 2015
by arrangement with Aram Publishing
English text edited by Joy Cowley
English edition © big & SMALL 2015

ISBN: 978-1-925233-70-4

Printed in Korea

Good Friends

Written by Ki-hwa Jang
Illustrated by Seung-min Oh
Edited by Joy Cowley

big & SMALL

Many animal friends work together.
Do you know who they are?

Animal mutualism is when two different species are in a relationship that benefits each other. It is a special type of interaction and a form of **symbiotic relationship**.

This bird is a honey guide.
The badger will follow it
to the bee's nest of honey.

The badger will open the nest
and leave some beeswax
for the bird.

🐝 Beeswax is a solid yellowish substance produced by bees for
making the cells in which they live. It is also called wax.

The badger and the honey guide
are very good friends.

The water buffalo has ticks
and flies that bite its back.
What can it do?

These oxpeckers eat the ticks and flies.
They think ticks and flies are delicious.
The water buffalo is glad
to have the birds on its back.

The water buffalo and the oxpeckers
are very good friends.

The ladybug comes to eat the aphids,
but the ants chase the ladybug away.

In return for protection,
the aphids give the ants
delicious sweet dew.

The ants and the aphids
are very good friends.

Sweet dew is excretion produced by aphids.
It is good for ants.

The wren calls the fish hawk.
It tells the fish hawk
that a snake is near its nest.

The fish hawk comes back
to protect its eggs.
It will also protect the eggs
in the wren's nest.

The wren and the fish hawk
are very good friends.

Some sharks are coming!
The goby has to hide
in the shrimp's nest.

When a predator fish approaches,
the goby warns the shrimp, so it too can hide.

The shrimp and the goby
are very good friends.

The sea anemone is on the hermit crab's back.
The hermit crab moves the sea anemone
to where they both find food.

Because the sea anemone has poison,
it protects the hermit crab from big fish.

The sea anemone and the hermit crab
are very good friends.

Despite their differences,
some animals are good friends.

The honey guide
and the badger

The oxpecker
and the water buffalo

The ants and the aphids

The fish hawk
and the wren

The shrimp
and the goby

The sea anemone
and the hermit crab

31

Good Friends

Animal mutualism is when two different species are in a relationship that benefits each other. It is a special type of interaction and a form of **symbiotic relationship**.

There are two other types of symbiotic relationship.
Commensalism describes a relationship where one species benefits, while the other is neither helped nor harmed. For example, barnacles attach to whales in order to get food and transportation.
Parasitism is when one species benefits at the expense of the other. For example, ticks attaching themselves to a dog. The tick benefits from eating the dog's blood, but the dog suffers from the loss of blood and nutrients and may get sick.

Let's think

How do relationships based on mutualism develop between animals?

Can you think of other examples of animal mutualism not in this book?

Is owning a pet a form of mutualism?

What are the names of the other types of symbiotic relationships?

Let's Do!

It's time to use your imagination. Think of some imaginary animals with different needs. Write down your imaginary animals and their needs. Think about which animals could have a relationship based on mutualism in order to get the things they need from each other.